Detail Design (DD), Front End (FE), Back End (BE) Design of Non-Tactical Deployable 20 GWh (17.20841 Kilo Ton TNT) Fusion Device – Trigger: Implosion Type Device

Muhammad Musaddique Ali Rafique

Published by Muhammad Musaddique Ali Rafique, 2023.

DETAIL DESIGN (DD), FRONT END (FE), BACK END (BE) DESIGN OF NON-TACTICAL DEPLOYABLE 20 GWH (17.20841 KILO TON TNT) FUSION DEVICE – TRIGGER: IMPLOSION TYPE DEVICE

First edition. March 9, 2023.

ISBN: 979-8231570164

Written by Muhammad Musaddique Ali Rafique.

Table of Contents

To my parents

Detail design (DD), front end (FE), back end (BE) design of non-tactical deployable 20 GWh (17.20841 kilo ton TNT) fusion device – trigger: implosion type device

Muhammad Musaddique Ali Rafique[1,2,3,4]

[1]Department of Fusion Engineering, Massachusetts Institute of Technology, Cambridge, MA 02139

[2]Department of Fusion Engineering, National Aeronautics and Space Administration, 39B, Merritt Island, Florida

[3]Department of Fusion Engineering, Eastern Engineering Solutions Inc. Cambridge, MA 02139

[4]Department of Fusion Engineering, Eastern Engineering Pty Ltd. Melbourne, VIC, AUSTRALIA

Abstract

A simple implosion type device is described. Its main parts, their detailed design, geometry (including placement and configuration), engineering and derivation from simple plane wave lens in earlier study is described. Types of lens, their configuration, truncated icosahedron, hexagonal, pentagonal, pyramidal cones, their arrangement, curve in polar co-ordinates and role is described. Pressure calculations are presented emphasizing explosive strength and bounds for criticality of fissile mass. Its detonation mechanism (arming) is also briefly explained. Various types of initiators and their operation is discussed. It is observed that simple mathematical principles play an important part.

Keywords: implosion, icosahedron, hexagonal, pentagonal, ice cream, explosive, initiator

1. Introduction

Low to medium range fusion devices have attracted a lot of attention recently due to their ability to use less (few grams [1]), simple, inexpensive, and abundant fuel (D_2 and T), have simple architecture and construction (using Co base superalloys [2]) with more new compact fission devices (triggers with new neutron initiators (electrical pulse)) and ability to produce large amounts of energy. These could be air dropped [3], canon fired or missile delivered with single or multiple entry and reentry vehicles (MIRVs). In this study, which is a continuation of previous studies [4, 5], an implosion type device (as trigger) is described. An implosion type fission device from spherical charge [6, 7] which forms the conglomerate of two stage modified Teller Ulm or Mike Ivy fusion device [1, 4, 5], is a process N which takes a spherical sub critical mass M of fissile material of radius R_1 as input, and divides it into a (n No. of finite) spherical *supercritical mass* of radius R_2 (Fig. I) [8]

Where

$$R_2 < R_1$$

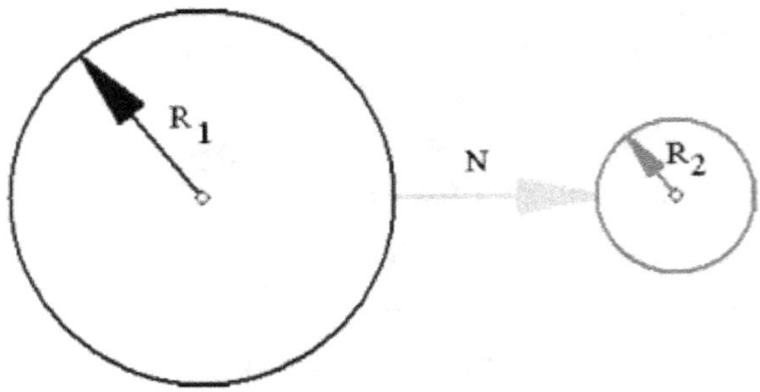
R1
N
R2

Fig. 1: Basic concept of an implosion type fission device [8]

In order for device to achieve said condition, a uniformly sufficient (integrated over whole surface) pressure is to be applied on the surface of sphere of fissile mass of radius R_1 such that mass (represented in the form of a sphere (left hand side in Fig. 1)) is compressed into a sphere of radius R_2 ($R_2 < R_1$). This may be represented in below Fig. 2

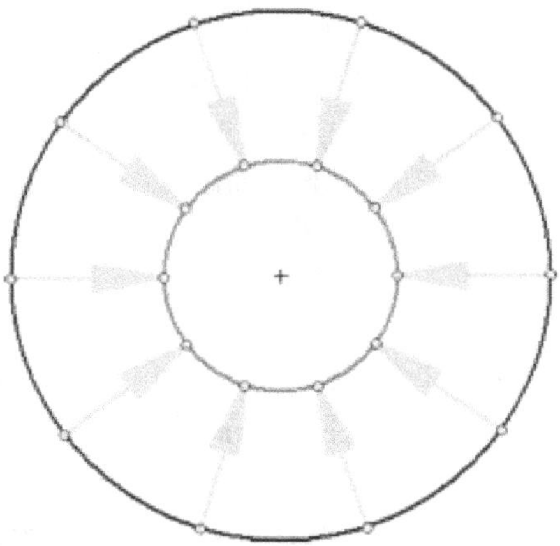

Fig. 2: Schematic of uniformly applied pressure integrated over whole surface to achieve sub critical mass [8]

Uniformly sufficient means that pressure must be applied in a spherical manner (integrated over whole surface), enough to reduce the radius from R_1 to R_2 with

$$R_2 \leq R_c < R_1$$

Where

R_c = critical radius (required) of fissile mass M.

R_c is a function of fissile element's (Uranium, Plutonium, Thorium) critical density (d_c), which in turn is a function of purity (> 99.9%) as achieved by extraction (extractive or chemical metallurgy [9-13]) and refining (enrichment e.g., of Uranium (Uranium Hexafluoride (UF_6)) [14, 15] by ultracentrifuges (gas centrifuges [16, 17]), diffusion, electromagnetic, aerodynamic, laser isotope, plasma separation and ion exchange (solvent exchange) processes).

To facilitate uniformity, a special enclosure is constructed, called a *shock – wave lens*. To facilitate sufficiency, a minimum pressure needs to be calculated which makes fissile mass M achieve a sufficient compression ratio. This minimum pressure is a function of radius of lens enclosure R_L and of the mass m_L of the conventional explosives in the lens enclosure used to deliver pressure against the fissile mass, M.

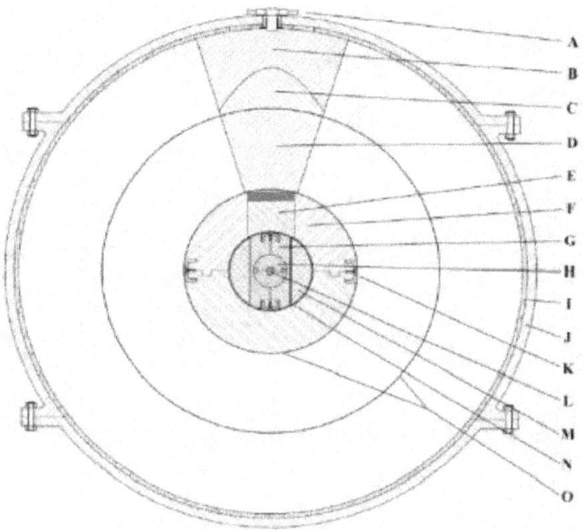

Fig. 3 a: A plane section of internal structure of implosion type device

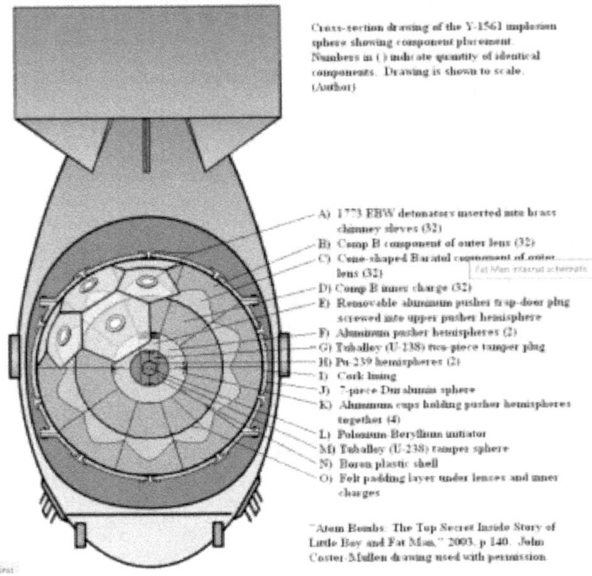

Fig. 3 b: A schematic of implosion type device, its internal and external structure

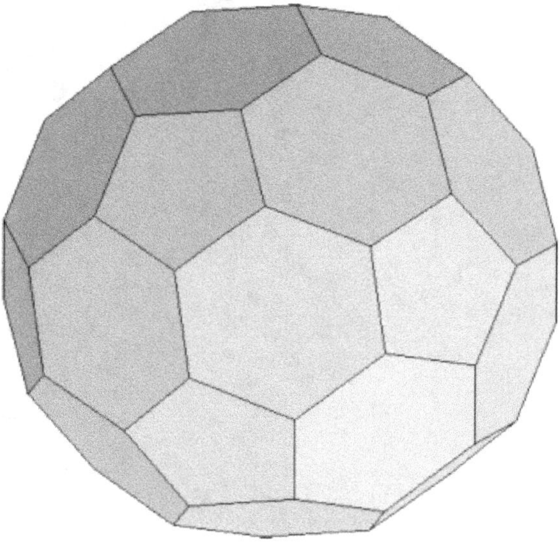

Fig. 4: Truncated Icosahedron lens apparatus for implosion type device

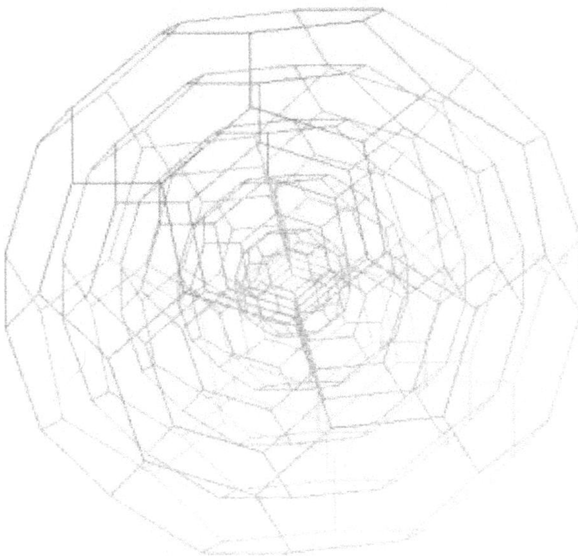

Fig. 5: Internal structure diagram for truncated icosahedron lenses

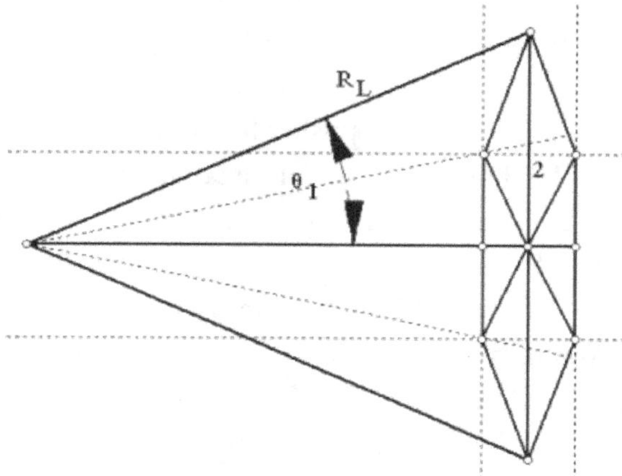

Fig. 6: Half apical angle for a hexagonal pyramidal cone

Both uniformity and sufficiency can be modelled using Maple [18] (a linear analytical model). Geometrical figures were constructed with Euclid.

1. Construction (main parts)

A plane section of internal structure of implosion type device is shown on figure 3 a and b. It consists of following main parts [6].

a. 1773 EBW detonator inserted into brass chimney sleeve.
b. Comp B component of outer polygonal lens
c. Cone shaped Baratol component of outer polygonal lens.

d. Comp B inner polygonal charge
e. Removable aluminum pusher trap door plug screwed into upper pusher hemisphere.
f. 18.5′ ø Aluminum pusher hemisphere
g. 4.375′ ø Tuballoy two-piece tamper plug
h. 3.62′ ø Pu-239 hemisphere with 2.75 diameter jet ring
i. 0.5′ thick cork lining.
j. 7 – piece Y – 1561 Duralumim sphere.
k. Aluminum cup holding pusher hemisphere together.
l. 0.8′ ø Polonium – Beryllium Urchin Initiator
m. 8.75′ ø Tuballoy tamper sphere
n. 9.0′ ø boron plastic shell
o. Felt padding layer under lenses and inner charge.

1. Uniformity: The shock wave lens enclosure

The shock wave lens, for simplicity called just *lens* consists of an almost spherical shell of radius RL, which focuses the explosive (shock) waves of the conventional explosive used to the center of the device, which contains the fissile material, called pit. This configuration (lens and pit) can have various models and geometries (such as truncated icosahedron, smaller icosahedron, frustum, pentahedron, hexahedron, pyramidal cone) explained below,

1. The Truncated Icosahedron Model (Nonlinear analytical model)

The lens usually is designed as a truncated icosahedron, which is almost spherical. The interior of this truncated icosahedron may be viewed as nested group of smaller truncated icosahedra resembling a honeycomb; therefore, each vertical lens element is either a hexahedron or pentahedron pyramidal cone / frustum of decreasing radius. On top and at the center of every hexagon and pentagon of the truncated icosahedron, there is an exploding-bridgewire detonator which initiates the explosion.

Because the lens enclosure is totally symmetric with respect to pit, each lens intersection with any plane normal to any radius will be a regular pentagon or hexagon whose symmetry group is either S_5 or S_6 respectively, therefore it can be modelled in two dimensions exactly, with the final model constructed by applying the principal symmetries of these two groups in three dimensions. Namely within each cone a complete rotation by 2π around the radial axis of the corresponding two-dimensional curve.

Inside each vertical lens cone there are two types of explosives used, a fast explosive with detonation speed $\vec{v}_1$, usually explosive compound B and a slow explosive with detonation speed $\vec{v}_2$, usually Baratol. Let R_A be the sub radius of the fast explosive prior to the focusing lens and R_B be the inner radius of the fast explosive after the focusing lens, with $R_A + R_B = R_L$

1. Apical Angles of Hexagonal and Pentagonal Pyramidal Cones

(A) If the vertices of the truncated icosahedron are placed on a sphere of radius $R_L = \sqrt{9\varnothing + 10}$ with $\varnothing = (1+\sqrt{5})/2$, then all the edges have length 2. Suppose circumradius of pentagons and hexagons are r_p and r_h respectively. Following equation satisfies the half apical angle θ_1 of the hexagonal cone and half apical angle θ_2 of the pentagonal cone respectively (Fig 6).

$$R_L \sin(\theta_1) = r_h$$

$$R_L \sin(\theta_2) = r_p$$

For the hexagonal cones, with all equilateral sub triangles, $r_h = 2$, and for the pentagonal cones, $r_p = \sqrt{50+10\sqrt{5}}/5$. Equations (A) becomes,

$$\theta_1 = \arcsin\left(\frac{4}{\sqrt{58+18\sqrt{5}}}\right)$$

$$\theta_2 = \arcsin\left(\frac{2\sqrt{50+10\sqrt{5}}}{5\sqrt{58+18\sqrt{5}}}\right)$$

From which we get apical angles of hexagonal and pentagonal cones as

$$\theta_1 = 2\theta_1 \sim 47.60036521 \text{ degrees}$$

$$\theta_2 = 2\theta_2 \sim 40.15350256 \text{ degrees}$$

These angles are invariant under R_L

1. Lens concept

Each cone has either five or six neighbors which have exact same behavior. All cones are synchronized in the explosion, therefore shock wave propagates only towards the pit and never cross boundaries, because they are cancelled by shock waves of neighboring cones.

Assume $(x(t), y(t))$ a point of shock wave front at time t after detonation at time $t = 0$. Clearly, $x(0) = 0$ and $y(0) = R_L$. Prior to focusing, $\rho_1(t) = \vec{v}_1 t$ is the wave front distance from detonator at $(0, R_L)$, as a function of time for fast explosive. Therefore, the shock wave for the fast explosive (Comp B) prior to focusing satisfy the equation:

$$x(t)^2 + (y(t) - R_L)^2 = (\vec{v}_1 t)^2 = \rho_1(t)^2 \qquad (1)$$

Which is an equation of circle. These shock waves are shown on Figure 7. The object of the lens is to focus the waves in the form shown on Fig 8. This can only happen if below equation is satisfied.

$$x(t)^2 + y(t)^2 = (R_L - \bar{v}_1 t)^2 = \rho_2(t)^2 \tag{2}$$

1. The hexagonal lens

The transition is achieved by adding a slow (less expensive) explosive with an appropriate shape Y1, which will modify the shock waves resulting from equation 2. This is a classical problem in wave mechanics first time investigated by Descartes for optical waves. Solution is called a Cartesian Oval. In this solution, a particular point exists, where hyperbola in place of a plane wave is achieved, which is one of degenerated cases of cartesian Oval. In order for Y1 to effectively work as lens between the detonator and pit, first one models a lens consisting of two media and then one modifies the design to minimize second medium, in this case slow explosive, the Baratol.

General equation of Cartesian Oval is

$$m\rho_1 + n\rho_2 = k \tag{3}$$

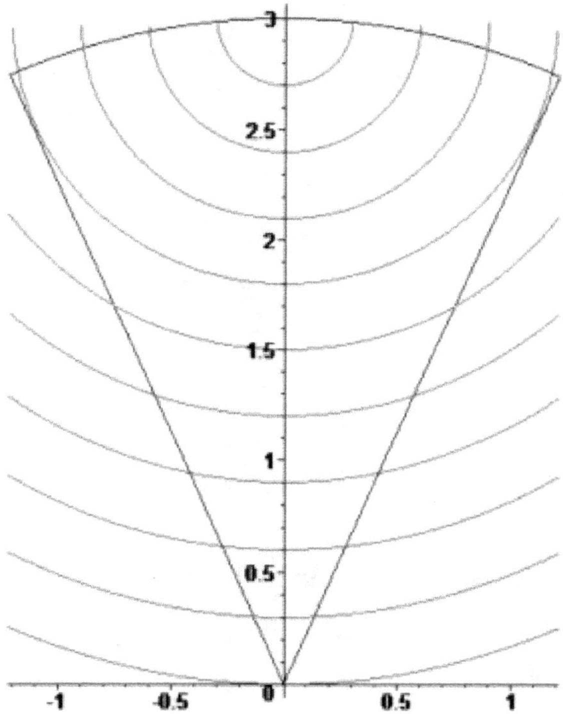

Fig. 7: Shock waves after detonation prior focusing.

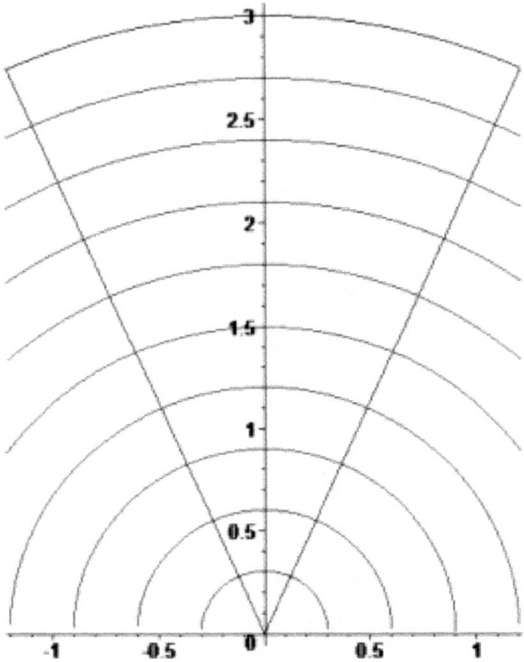
3
2.5
2
1.5
1
0.5
0
-1
-0.5
0.5
1

Fig. 8: Shock waves after focusing.

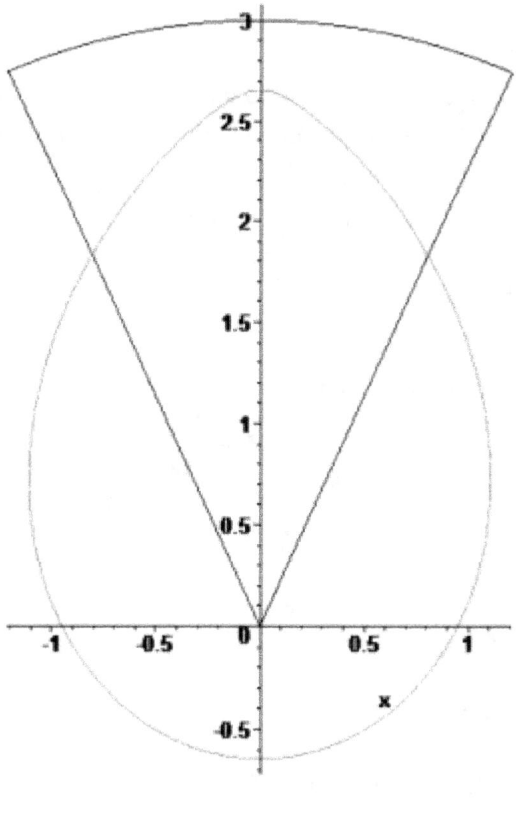

Fig. 9: Cartesian Oval Lens for two foci, at (0, R_L) and (0, 0), with R_L = 3 and r ~ 0.11 for hexagonal lens.

For a vertical cone model (volume with disc method revolving around x or y axis (described previously [4, 5] and elsewhere)) with detonator at (0, R_L) and pit at (0, 0) and in view of equations (1) and (2), equation (3) becomes,

$$m\sqrt{x^2 + (y - R_L)^2} + n\sqrt{x^2 + y^2} = k$$

Suppose

$\rho_1 = rR_L$, with $0 < r < 1$

we have

$\rho_2 = (1 - r)R_L$

and then

$$k = m\rho_1 + n\rho_2$$

Putting values

$$k = mrR_L + n(1 - r)R_L$$

Rearranging

$k = (mr + n(1 - r))R_L$

For generality, m = 1.

Note that, $n/m = \vec{v}_1/\vec{v}_2$, hence for two explosives, Comp B and Baratol, which have $\vec{v}_1$= 8050 m / sec and $\vec{v}_2$= 4900 m / sec respectively, n = 23 / 14

In view of the above and for R_L = 3 m (3) becomes

$$m\sqrt{x^2 + y^2 - 6y + 9} + \frac{23}{14}\sqrt{x^2 + y^2} = \frac{-27r + 69}{14} \qquad (4)$$

Assume solution of equation 4 is y. It must be radially symmetric as it transitions a radially symmetric shock wave into another. Resultant wave is shown in Fig 9 above.

From equations 2, 3 and 4, it is obvious that shape of Y is a function of the speed of explosive used n / m, and hence of R_A and finally of R_B. Different explosives and radii R_A and R_B will require different shaped Y. That's why in practical terms the exact shape is machined using iteration, by checking $y(r,x_1)$ and $R_B\cos(\theta_1)$ for equality, with $x_1 \in [0.1]$ a solution of the equation $y(r,x) = x\tan(\pi/2-\theta_1)$, which must hold at the edge of the cone. For a given R_B one choses an r which closely matches R_B. The iteration is complete when r satisfies the equation,

$$y(r,x_1) = R_B\cos\theta_1 \qquad (5)$$

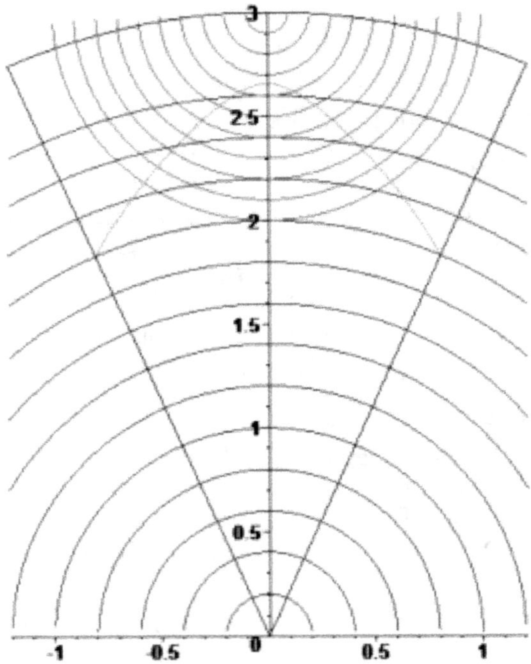
3
2.5
2
1.5
1
0.5
-1
-0.5
0
0.5
1

Fig. 10: Shock wave before focusing (red), after focusing (blue) and transition curve *y*(*x*,*y*) (gold), with $R_A = 1$m, $R_B = 2$m, r ~ 0.11 and n = 23/14 for hexagonal lens.

This all is non-linear analytical modeling based on iteration and trail an error. This is laborious and tedious process usually requiring careful CNC machining (not 3D printing) described elsewhere. In this work, figures are created after two iterations of x_1 and r to satisfy $R_B \sim 2R_L / 3$. Once the shock wave has been focused below the transition surface of hexagonal and pentagonal cones to a base extent of r_h and r_p respectively, there is no need to add more Baratol slow explosive below R_B. Therefore, from a practical construction standpoint, Baratol explosive can only be an additional slice (called ice cream cone) added on top of the fast Compound B explosive inside every cone at radius R_B as shown in below figure (Figure 11). For a moment, ignoring tamper, pusher, and other elements (e.g. Po-Be neutron initiator [19, 20]) which lie close to pit (old designs – new designs employ Uranium Deuteride (UD_3), electrically triggered pulsed neutron emitters and others e.g. Heavy ions [21], Eulerian mathematical method [22], Fluids [23], D-T gas compression [24]) (F, G, H in Fig 3) [25], it can be seen that internal Compound B explosive can be a solid sphere of radius R_B. As described earlier, by symmetry of the group S_6, the required surface Y will then just be curve y of equation 4 revolved around y axis by 2π radius as shown in Figure 9. This surface along with hexagonal cone is shown in Figure 12 below.

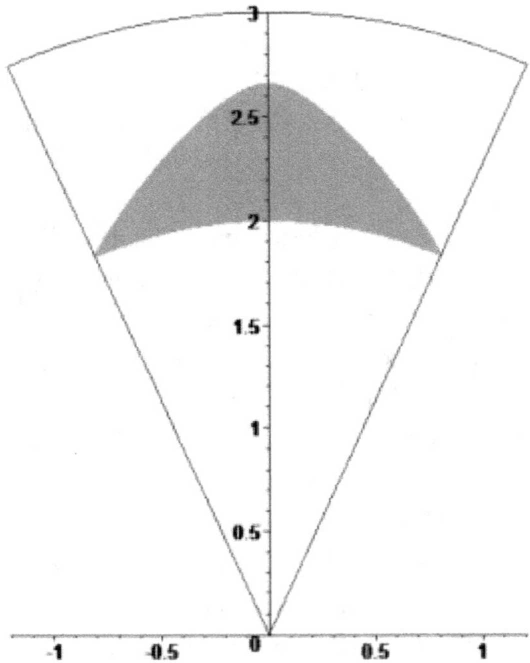
3
2.5
2
1.5
1
0.5
-1
-0.5
0
0.5
1

Fig. 11: Ice cream cone with Baratol explosive at radius R_B = 2m for hexagonal lens

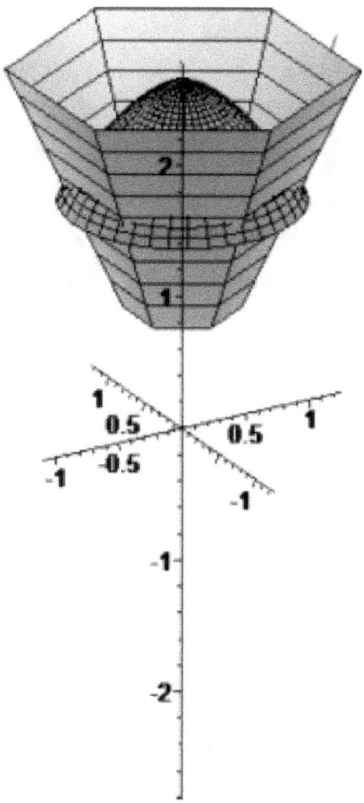

Fig. 12: Baratol explosive surface Y (*r*, *x*) at radius R_B = 2m for hexagonal lens

1. Pentagonal Lens

The analysis for pentagonal cone is also similar. In this case, ϑ_1 is replaced by ϑ_2 and S_6 by S_5. It is imperative that both hexagonal and pentagonal lenses must be synchronized at (not after) detonation. This must be accounted in predesign, construction, iteration and final design, assembly, and arming. Any mismatch will result in non-functioning or mal functioning of device. This can only be achieved if high explosive (Compound B) at radius R_B must ignite at the same time (t_B) as it ignited for hexagonal lens.

$$R_B = 2R_L/3 \quad (6)$$

$$t_B = \frac{rR_L}{\vec{v}_1} + \frac{\left(\frac{1}{3}-r\right)R_L}{\vec{v}_2} \quad (7)$$

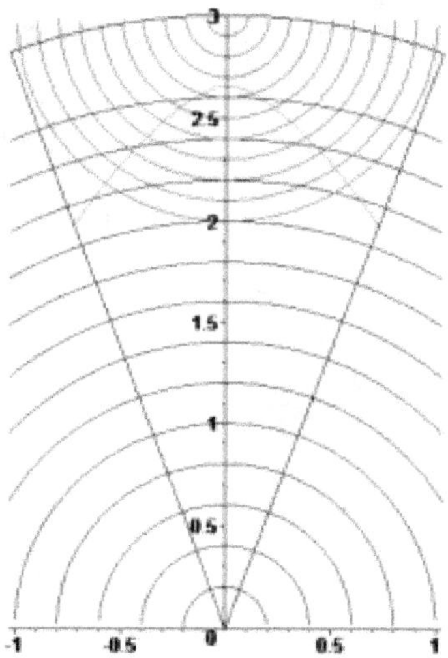

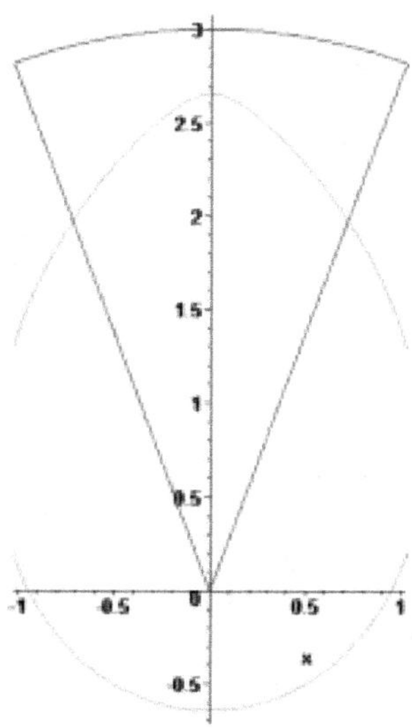

Therefore, same r and hence same Cartesian Oval can be used for both hexagonal and pentagonal lenses with adjustment of height only in pentagonal lens. This seems intuitively correct as well as Cartesian Oval as a locus is invariant under different cone boundaries. The results are shown in below figures. (Figure 13 a – d).

(b
)

(a
)

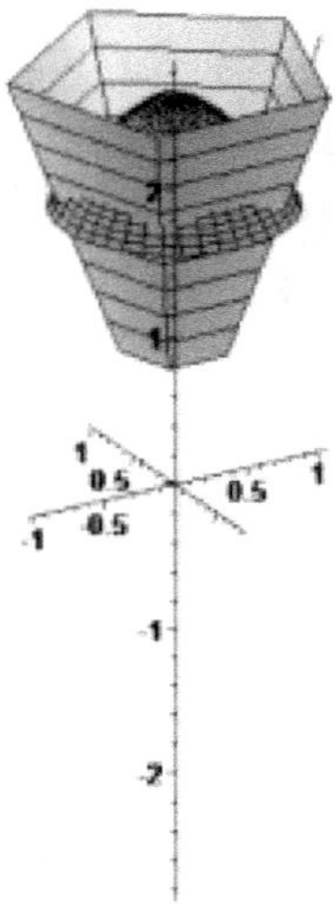
2
1
1
0.5
0.5
1
-1
-0.5
-1
-2

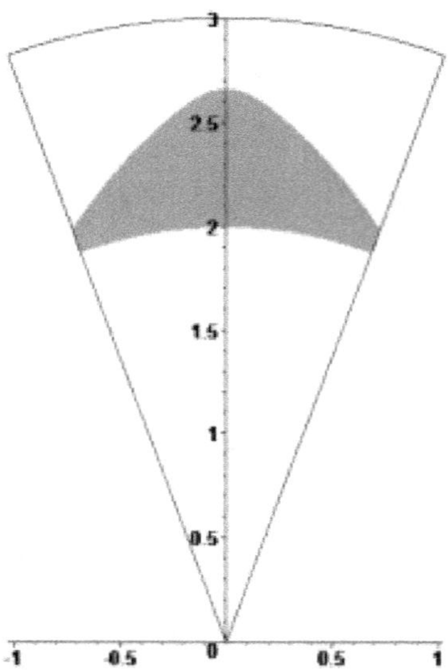
3
2.5
2
1.5
1
0.5
-1
-0.5
0
0.5
1

(c
)

(d
)

Fig. 13 (a – d): Results of pentagonal lens

1. The Lens Curve in Polar Coordinates

Substitute x = Rcos(θ) and y = Rsin(θ) in equation 4 and solve for R in terms of θ, the resultant will be a lens curve in polar coordinates.

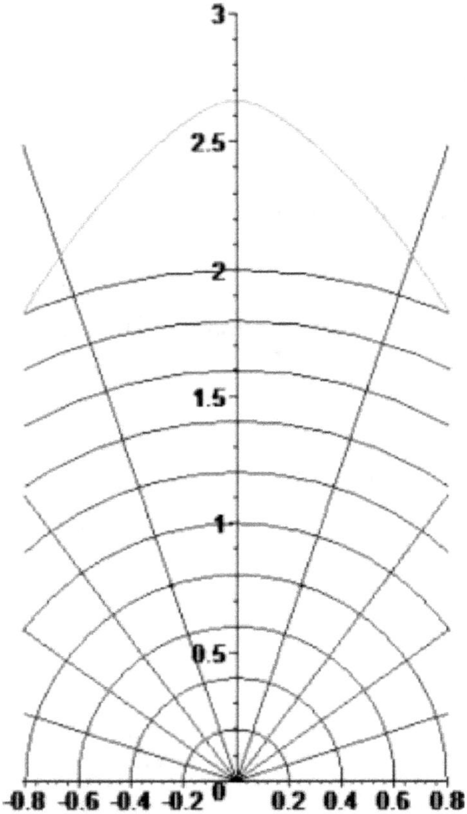
3
2.5
2
1.5
1
0.5
0
-0.8 -0.6 -0.4 -0.2
0.2 0.4 0.6 0.8

Fig. 14: Transition curve y = R(θ) in Polar coordinates

As a quadratic will result, there will be no closed form expression in most cases. A polar form is preferable because polar coordinates are more suitable for machining radially symmetric surfaces and numerical polar form results. For above example, a similar equation (polar equation for transition surfaces) to five decimal places and two iterations for both (hexagonal and pentagonal) lenses may be written as

$$R(\theta) = -1.76576\sin(\theta) + 4.55343 - 0.84084\times10^{-10}\cdot\sqrt{0.441\times10^{21}\sin(\theta)^2 - 0.22744\times10^{22}\sin(\theta) + 0.18358\times10^{22}} \quad (8)$$

Where

$\pi/2 - \theta_1 \leq \theta \leq \pi/2 + \theta_1$ for hexagonal lens and

$\pi/2 - \theta_2 \leq \theta \leq \pi/2 + \theta_2$ for pentagonal lens

Resulting shape is shown in figure 14 above.

1. Machining of Baratol

Machining of Baratol may be performed by using above procedure and equations. To do this, following steps are performed,

a. Set r = r1 in equation 4
b. Consider incremental approximations θ_j, where $j \in \{0,1,2,\ldots,N\}$ on equation 8

The required approximate radial distances to the surface will be given as $R(\theta_j)$, and j as above in two dimensions. In this scenario, when N = total machining steps n interval [$\frac{\pi}{2} - \theta_1, \pi/2$] and tolerance $\Delta R \leq 10^{-3}m$ = 1mm, resulting machining steps would be $N \geq 57$ which corresponds to $\Delta\theta \leq 0.41754°$ / step size. Resulting lenses are shown below (Figure 15 a and b). The higher the Baratol lenses are positioned inside the corresponding pentagonal or hexagonal cone, the greater the delivered force of Compound B explosive would be.

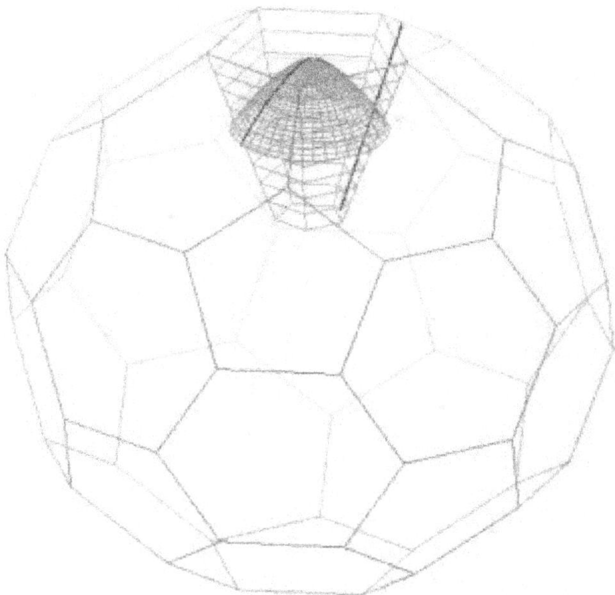

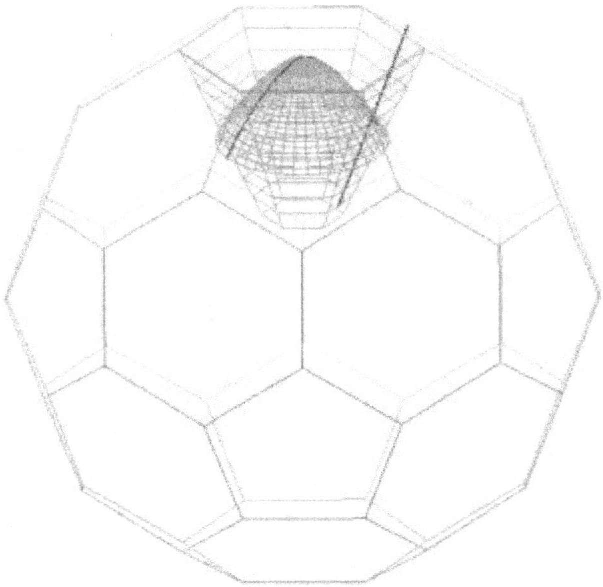

(b
)

(a)

Fig. 15: Lenses (a) Hexagonal, (b) Pentagonal

1. **Calculation of Pressure**

This section describes how much pressure may be generated by detonation of a device. As described earlier, TNT is used as common unit of measurement [1, 4, 5].

1. **Explosive Strength**

As discussed earlier, radius of lens is R_L, and radius of fissionable mass is R_F, then total volume of lens is

$$V_L = \frac{4}{3}\pi(R_L^3 - R_F^3) \quad (9)$$

Density (ρ) of TNT ~ Comp B = 1.65 g / cm^3, total mass

$$m_L = \rho . V_L \quad (10)$$

Explosive strength of TNT (s) = 2.72 x 10^6 J / Kg, total explosive strength

$$S_L = s . m_L \quad (11)$$

Suppose volume of fissile element = V_F, then total pressure P_L delivered to fissile element by lens

$$P_L = \frac{S_L}{V_F}$$

Note: MKS (or SI) system of units is used

1. Radial bounds for criticality of fissile mass

Plutonium's bulk modulus K_{Pa} and can be calculated by its young and shear modulus E_{Pa} and G_{Pa} respectively.

$$K = \frac{EG}{3(3G-E)} \quad (12)$$

Note: This equation approximates Irons bulk modulus quite well (giving K_{Fe} 164 GPa vs 170 GPa experimental). However, it is deceiving. K_{Pu} = 41.69 GPa versus a reported experimental K_{Pu} = 55 GPa. With safety factor, K_{Fe} = 160 GPa and K_{Pu} = 60 GPa

For Plutonium 239, its density is to be increased only by a factor of 2 for triggering and sustainability of fission. Although much higher densities are still possible with advanced designs. If initial and final densities are ρ_1 and ρ_2 corresponding to initial and final volumes v_1 and v_2, then,

$$\rho_2 \geq 2\rho_1$$

$$\frac{m}{V_2} \geq 2\frac{m}{V_1} \quad (1$$

$$V_2 \leq \frac{1}{2}V_1$$

The above will require a pressure increase of at least K_{Pu} / 2. From equation 11, we get,

$$P_L \geq \frac{K_{Pu}}{2}$$

Substituting in 12

$$\frac{s\rho(R_L^3 - R_F^3)}{R_F^3} \geq \frac{K_{Pu}}{2} \quad (13)$$

If appropriate values are inserted, it can generate radius and mass required for explosion. This leads to design of various devices described below.

1. Case 1: RF = 5 cm

Rearranging above equation (13) and solving for RL

$$(R_L^3 - R_F^3) \geq \frac{R_F^3 K_{Pu}}{2s\rho} \quad (14)$$

$$R_L^3 \geq \frac{R_F^3 K_{Pu}}{2s\rho} + R_F^3 \quad (15)$$

$$R_L \geq \sqrt[3]{\frac{R_F^3 K_{Pu}}{2s\rho} + R_F^3} \quad (16)$$

$$R_L \geq \sqrt[3]{\left(1 + \frac{K_{Pu}}{2s\rho}\right) R_F^3} \quad (18)$$

Substituting, s = 2.72 x $10^6 \, J / Kg$, ρ = 1650 Kg / m^3, RF = 5 cm = 5 / 100 m, KPu = 60 GPa in above

$$R_L \geq 9.88 \text{ cm} \quad (19)$$

Hence for this case,

$$\Delta R = R_L - R_F \geq 9.88 - 5 \text{ cm} = 4.88 \text{ cm} \quad (20)$$

This leads to

$$\Delta m_L = \rho \Delta V$$

From equation 9

$$\Delta m_L = \rho V = \rho \frac{4}{3}\pi\left(R_L^3 - R_F^3\right)$$

Putting values and solving

$$\Delta m_L = 5.71215 \text{ Kg TNT}$$

1. **Case 2: R_F = 2.5 cm**

From equation (18)

$$R_L \geq \sqrt[3]{\left(1 + \frac{K_{Pu}}{2s\rho}\right) R_F^3} \quad (2$$

Substituting, s = 2.72 x 10^6 *J / Kg*, ρ = 1650 *Kg / m*^{3}, R_F = 2.5 cm = 2.5 / 100 m, K_{Pu} = 60 GPa in above

$$R_L \geq 4.93 \text{ cm} \quad (2$$

Hence for this case,

$$\Delta R = R_L - R_F \geq 4.93 - 2.5 = 2.43 \text{ cm} \quad (23)$$

This leads to

$$\Delta m_L = \rho \Delta V$$

From equation 9

$$\Delta m_L = \rho V = \rho \frac{4}{3}\pi\left(R_L^3 - R_F^3\right)$$

Putting values and solving

$$\Delta m_L = 0.719800 \text{ Kg TNT}$$

1. **Case 3: R_F = 7.5 cm**

From equation (18)

$$R_L \geq \sqrt[3]{\left(1 + \frac{K_{Pu}}{2s\rho}\right) R_F^3} \quad (24)$$

Substituting, s = 2.72 x 10^6 *J / Kg*, ρ = 1650 *Kg / m*^{3}, R_F = 7.5 cm = 7.5 / 100 m, K_{Pu} = 60 GPa in above

$$R_L \geq 75.21 \text{ cm} \quad (25)$$

Hence for this case,

$$\Delta R = R_L - R_F \geq 75.21 - 7.5 = 67.71 \text{ cm} \quad (26)$$

This leads to

$$\Delta m_L = \rho \Delta V$$

From equation 9

$$\Delta m_L = \rho V = \rho \frac{4}{3} \pi \left(R_L^3 - R_F^3\right)$$

Putting values and solving

$$\Delta m_L = 2935.9470 \text{ Kg TNT}$$

If a tamper and pusher which are made up of Tuballoy and Aluminum respectively, are used as shown in above figures, it suffices to take $R_F' = R_P$ where R_P = radius of pusher and $K_N \sim$ max {K_{Pu}, K_U, K_{Al}} = 100 GPa, because this is space which needs to be compressed altogether. In this case, RP = 23.5 cm and solving above equation (24) with same parameters, gives $\Delta R \geq 54.13$ cm and $\Delta m_L \geq 988.9$ kg of TNT. Note: this is far less than quantity obtained in case 3 above. In this case (special case), minimum diameter of device is D_N = 2 x (R_P + ΔR) = 2 x (23.5 + 54.13) ~ 1.55 m (close to size of Fat man ~ 1.4 m). For this D_N and for r ~ 0.11386 and after two iterations for hexagonal lens, results are

$R_A = 25.87666$ cm and $R_B = 51.75333$ cm

Which is approximately twice the radius of R_P of pusher F shown on figure. To increase the devices detonation pressure even further, one can move Baratol lens close to top, which may be obtainable by using smaller value of r in iteration as shown above in case 1.

1. Assembly and detonation mechanism (arming)

Devices many be preassembled and stored or deployed (present case) and armed at the point when required to be detonated. Modern detonation mechanisms are simpler (e.g. externally, electrically operated pulse neutron generators), mobile, agile and robust which ease out detonations in minutes.

1. Assembly

Assembly of the device is carful process carried out of following steps,

a. 32 separate frustums (Figure 6) are created (20 hexagonal and 12 pentagonal ones) as shown in Figure 15 and 16 which contain the lenses.
b. The frustums are then assembled on top of the pusher sphere completing the truncated icosahedron (Fig 4 and 5)
c. One hexagonal cone is left uncovered, and its contents can be removed based on diagram of figure 3.

This includes all the elements inside the cone shown on diagram.

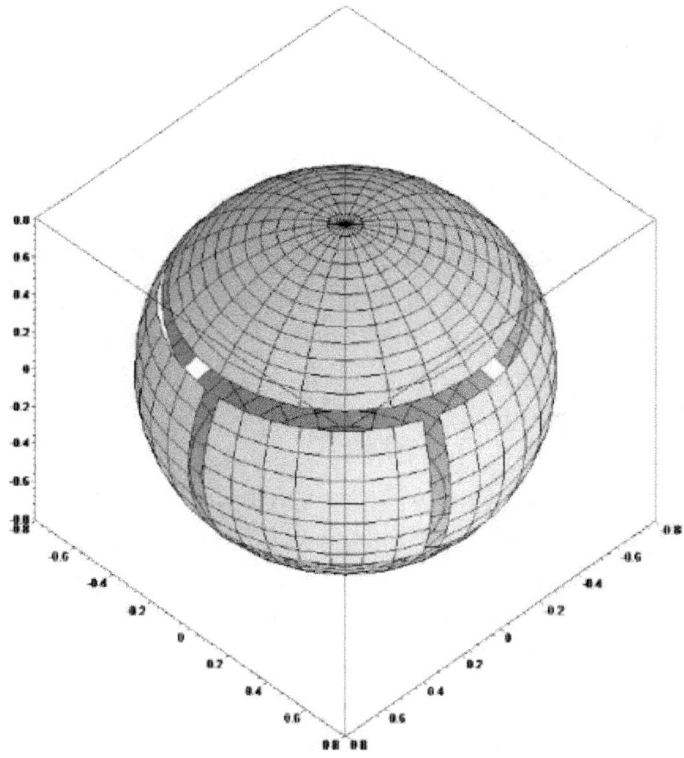
0.8
0.6
0.4
0.2
0
-0.2
-0.4
-0.6
-0.8

Fig. 16: External Duralumin sphere assembly of device shown in Fig 3 a, b

1. Arming

Arming sequence and steps are described below,

a. First neutron generator (e.g. Po-Be urchin neutron initiator [19, 20]) or Uranium Deuteride (UD_3) or electrically triggered pulsed neutron emitter or others (section 2.4) (L)) is placed at the center of two plutonium hemispheres (H)
b. Join the two hemispheres.
c. Inserts the resulting sphere into one of the two semi cylinders of Tuballoy two piece tamper plug (G)
d. Join the two pieces together and screw the resulting cylinder into the removable aluminum pusher trap door plug (E).
e. The resulting segment is then lowered into the actual pit and screwed against the pusher element (F).
f. Last hexagonal lens is put into place
g. Last 7 piece, Y – 1561 Duralumin (top) cap (J) is placed on the resulting sphere
h. Last 1773 EBW detonator (A) inserted into brass chimney sleeve.

This arming is carefully controlled and carried out and has been considerably eased out by the invention of modern optics, metrology, measurement tools and diagnostics.

1. Conclusion

Fission device which triggers a fusion device is described with optimum actual geometrical values. A standard implosion type device consisting of a truncated icosahedron, and hexagonal cones in it is explained with emphasis on shock wave focusing using Cartesian Oval. Design, arrangement and positioning of cones, high explosives, low explosives (Baratol), initiators, and pit are described in simpler manner using 3D non-linear analytical solutions. Construction of device and need for iteration in machining is emphasized. Few new methods for neutron generators or initiators are also described citing their significance, design, and advantage over the old ones. Modern metrology and diagnostics play an important role in arming and detonation of small to medium output yield devices.

1. References

1. Rafique, M.M.A., *Design of Non-Tactical Deployable 20 Gwh (17.20841 Kilo Ton TNT) Fusion Device-Energy Basis.* 2023.

2. Rafique, M.M.A., *Plasma facing material by self-interstitial solid solution strengthening–problem, proposition, and a solution.* 2023.

3. Rafique, M.M.A., *Materials for outer shell of 1.170 GWh (1.00669 kilo ton TNT) fusion device-weight basis.* 2023.

4. Rafique, M., *Detail design (DD), front end (FE), back end (BE) design of non-tactical deployable 20 GWh (17.20841 kilo ton TNT) plane wave lens configuration fusion device.* 2022.

5. Rafique, M., *Detail design (DD), front end (FE), back end (BE) design of non-tactical deployable 20 GWh (17.20841 kilo ton TNT) plane wave lens configuration fusion device.* 2022.

6. Popov, N., et al., *Thermonuclear fusion in the explosion of a spherical charge (the problem of a gas-dynamic thermonuclear fusion)*. Physics-Uspekhi, 2009. **51**: p. 1047.

7. Popov, N.A., et al., *Thermonuclear fusion in the explosion of a spherical charge (the problem of a gas-dynamic thermonuclear fusion)*. Physics-Uspekhi, 2008. **51**(10): p. 1047.

8. Galidakis, I. and Y. Galidakis, *On Certain Design Aspects of the Implosion-Type Nuclear Weapon (ver. 1.1)*. 2009.

9. Parker, R.H. and D.W. Hopkins, *An Introduction to Chemical Metallurgy: International Series on Materials Science and Technology*. 2016: Elsevier Science.

10. Moore, J.J., *Chemical Metallurgy*. 2013: Elsevier Science.

11. Merritt, R.C., *EXTRACTIVE METALLURGY OF URANIUM*. 1971, Country unknown/Code not available.

12. Merritt, R.C., *Extractive metallurgy of uranium, The*. Colorado School of Mines. Arthur Lakes Library.

13. Du Preez, J.G.H., et al., *The chemistry of uranium. Part XXVIII. The development of a combined gold and uranium leach of randfontein ore*. Hydrometallurgy, 1981. **6**(3): p. 203-218.

14. Edwards, C.R. and A.J. Oliver, *Uranium processing: A review of current methods and technology*. JOM, 2000. **52**(9): p. 12-20.

15. Villani, S., *Uranium enrichment*. 1979, United States: Springer-Verlag,New York, NY.

16. European, C., et al., *Uranium enrichment by gas centrifuge*. 1990: Publications Office.

17. Heriot, I.D., *Uranium enrichment by gas centrifuge*. 1988: Commission of the European Communities (CEC). p. 141.

18. Redfern, D.A. *The Maple Handbook: Maple V Release 3*. 1995.

19. Rhodes, R., *The Making of the Atomic Bomb*. 2012: Simon & Schuster.

20. *Polonium-210 Program*. 1969: United States. p. Medium: ED; Size: 49 pages.

21. Tahir, N.A., et al., *Shock compression of condensed matter using intense beams of energetic heavy ions.* Physical Review E, 2000. **61**(2): p. 1975-1980.

22. Yoh, J.J. and K.-h. Kim, *Shock compression of condensed matter using Eulerian multimaterial method: Applications to multidimensional shocks, deflagration, detonation, and laser ablation.* Journal of Applied Physics, 2008. **103**(11): p. 113507.

23. Kim, K.-h. and J.J. Yoh, *Shock compression of condensed matter using multimaterial reactive ghost fluid method.* Journal of Mathematical Physics, 2008. **49**(4): p. 043511.

24. Rusov, V., et al., *Impulse source of high energy neutrons emitted by fusion reactions after compression of D-T gas by cumulative detonation waves.* Journal of Physical Studies, 2017. **23**.

25. Arrigo, L.M., J.I. Friese, and L.A. Metz, *Fabrication and Characterization of Plutonium Targets for Irradiation in the Flattop Critical Assembly*. 2020: United States. p. Medium: ED.

Don't miss out!

Visit the website below and you can sign up to receive emails whenever Muhammad Musaddique Ali Rafique publishes a new book. There's no charge and no obligation.

https://books2read.com/r/B-A-KRTHB-CGCCD

BOOKS 2 READ

Connecting independent readers to independent writers.

Did you love *Detail Design (DD), Front End (FE), Back End (BE) Design of Non-Tactical Deployable 20 GWh (17.20841 Kilo Ton TNT) Fusion Device – Trigger: Implosion Type Device*? Then you should read *Fusion Materials Semi Decadal Books of Abstracts*[1] by Muhammad Musaddique Ali Rafique!

[2]

This is compendium of book of abstracts published during last five years of symposium Fusion Materials. It collects and presents all developments till date at one platform for a ready reference.

Read more at https://alirafique.com/.

---

1. https://books2read.com/u/bPMlWR
2. https://books2read.com/u/bPMlWR

Also by Muhammad Musaddique Ali Rafique

Fusion Materials

Fusion Materials Semi Decadal Books of Abstracts

Fusion Device Design – Focusing of Shock Waves in Air and Gases. Part I: Formation and Analysis in One Dimension

Fusion Materials III - Book of Abstracts revised

Space Materials

Human Space Fight In Unpowered, Unbalanced, Inefficient Single Barrel Multistage Pencil Rockets – SpaceX Crew 1, Revisited

Materials for Human Space Flight III - Book of Abstracts Revised

Standalone

Composite Materials - Techniques and Procedures

Composite Materials - Resin Chemistry and Rheology

EBSD microstructure mapping of bulk metallic glass matrix composites

Inventory Management – A Model for advance strategic procurement

Laser additive manufacturing of Inoculated and non - Inoculated metallic glass matrix composites

Metallic glass matrix composites by inoculation and their laser additive manufacturing I

Metallic glass matrix composites by inoculation and their laser additive manufacturing II

Metallic glass matrix composites by inoculation and their laser additive manufacturing III

New DNA based hybrid catalysts

Strategies for the improvement of transportation system for efficient supply of goods

Heat treatment of piston rings

Bulk metallic glasses and their composites: Additive Manufacturing ad Modelling and Simulation Vol II

Modelling and simulation of microstructural evolution in Zr based Bulk Metallic Glass Matrix Composites (BMGMC) in Additive Manufacturing – a proposal, opinion and prospect.

Bulk Metallic Glasses and Their Composites — Methods, Prospects and Outcomes

Modeling and simulation of microstructural evolution in Zr based bulk metallic glass matrix composites during solidification

Orientation relationships in bulk metallic glass matrix composites

Orientation relationships in microstructure and calculation of pair distribution function (PDF) for additive manufacturing of Zr47.5Cu45.5Al5Co2 metallic glass matrix composites

Orientation relationships in inoculated and non inoculated Zr based bulk metallic glass matrix composites for laser additive manufacturing
Microstructural evolution in Zr based Bulk Metallic Glass Matrix Composites (BMGMC) in Additive Manufacturing - Modeling and simulation
Microstructural evolution in Zr based Bulk Metallic Glass Matrix Composites (BMGMC) in Additive Manufacturing - Modeling and simulation II
Microstructural evolution in Zr based Bulk Metallic Glass Matrix Composites (BMGMC) in Additive Manufacturing - Modeling and simulation III
Microstructural evolution in Zr based Bulk Metallic Glass Matrix Composites (BMGMC) in Additive Manufacturing - Modeling and simulation IV
Modelling and simulation of microstructure evolution during laser additive manufacturing
Orientation relationships (OR) in microstructure and calculation of pair distribution function (PDF) for additive manufacturing (AM) of Zr47.5Cu45.5Al5Co2 metallic glass matrix composites (MGMCs)
Extra-terrestrial Mapping Using Artificial Intelligence – a NASA Mars Rover Study
Detail Design (DD), Front End (FE), Back End (BE) Design of Non-Tactical Deployable 20 GWh (17.20841 Kilo Ton TNT) Fusion Device – Trigger: Implosion Type Device
Design Of Implosion Type Device (Trigger For Fusion Device)
Materials for Human Space Flight IV - Book of Abstracts
Corrosion and Microstructure Evolution of Ultrasonic (UaFSW), Induction (i-FSW) and Laser Heating Assisted

Macro and Micro (µFSW) Dissimilar Friction Stir Welding and Friction Stir Spot Welding (FSSW)
321 SS And Co Base Superalloys As Plasma Facing Material
Point Source Explosion in Air
Flexural Properties, Durability and Fatigue Life Evaluation (S–N Curves and Constant-Life Diagrams) of Aligned Plant Fibre Civil Composites with Various Configurations and Architecture.
Metallic Glass Matrix Composites VII - Book of Abstracts
Materials for Human Space Flight V - Book of Abstracts
Fusion Materials V - Book of Abstracts
Fusion Materials IV - Book of Abstracts Revised

Watch for more at https://alirafique.com/.

About the Author

Dr. Muhammad Musaddique Ali Rafique is Full Professor, Vice Chancellor, Head of Department (Department of Fusion Engineering and Department of Space Engineering) and Founding Director (Master of Science in Engineering (M.Sc.), Fusion Engineering and Master of Science in Engineering (M.Sc.), Astronautics and Manned propulsion) at Massachusetts Institute of Technology and President, Chief Executive Officer and Senior Principal Engineer at Eastern Engineering Solutions Inc.. He is also Senior Principal Engineer at Departments of Space Engineering and Fusion Engineering, National Aeronautics and Space Administration (NASA), Government of the United States of America.He is author of more than 162 peer reviewed and non peer reviewed journal articles, preprints, abstracts, conference papers, abstracted and non abstracted reports, book chapters, monographs, edited volumes and books. He has wealth of industrial experience in fields of rocket propulsion, energetics, fusion reactor design, device design, alloy development, inspection and testing, manufacturing and astronautics.He has myriad of research interest in metallurgy, extractive metallurgy (pyro, electro and hydrometallurgy), chemical metallurgy, rare earths, refining, cell design, operation, hydrogen metallurgy, plasma metallurgy, electric arc furnace (EAF) and its varients, AOD, VOD, fusion science and engineering, plasma facing materials, irradiation damage, ion irradiation, embrittlement, physical metallurgy, mechanical metallurgy, elsticity, plasticity, damage mechanisms, strengthening mechanisms, phase formation and transformations, Additive manufacturing, 3D printing, welding and joining, heat treatment, post weld heat treatment (PWHT), mechanical testing, fatigue, its types (e.g. fretting fatigue, corrosion fatigue, thermal fatigue, thermocyclic fatigue, bending fatigue), creep, its types, corrosion, its types (e.g. galvanic corrosion, pitting, crevice corrosion, stress corrosion cracking (SCC), intergranular corrosion, high temperature corrosion, fretting corrosion, fatigue corrosion, selective leaching, microbial corrosion), materials characterization (SEM, EDS, WDS, EBSD, FIB, AFM, APT, TEM, SRIM, CT, NDT), catalysts, plasma chracterisation, diagnostics, fusion device and reactor design and engineering, metallic glasses and their composites, high entropy alloys (HEAs), modeling and simulation (Part scale using ABAQUS CAE, Ansys (WB and APDL), MATLAB and FORTRAN, determ...

Read more at https://alirafique.com/.

About the Publisher

Dr. Muhammad Musaddique ali Rafique is a professional engineer. He is President, Chief Executive Officer and Senior Principal Engineer at Eastern Engineering Solutions LLC where is publishing under self publishing model. His work include areas such as fusion engineering, space engineering and modifying and refining existing approaches and techniques to reach at extraterrestrial (robotic and human).

Read more at alirafique.com.

www.ingramcontent.com/pod-product-compliance
Lightning Source LLC
LaVergne TN
LVHW022319200926
848481LV00004B/1374
* 9 7 9 8 2 3 1 5 7 0 1 6 4 *